The Trinity Balance in Reality

The Trinity Balance in Reality

A First Principle Way to Understand Life, the Universe, and Awe

Michael C. Longo

"Life exists within the union of two complementary parts."

3H Company/Balance Books

Lula, Georgia, USA

ISBN: 978-0-9676704-5-4

Dedication

To those who want to become enlightened, and live in paradise.

Authors note

Truth is often hidden in plain sight, obscured by noise, and half-understood ideas. This book offers a way to see through that confusion.

Within these pages, you will discover the Trinity Balance in Reality, the simple, three-part pattern that underlies all things. By learning to recognize it, you gain a tool for seeing reality as it is, not as others claim it to be.

The Trinity Balance in Reality is not theory or philosophy; it is the grammar of existence. It teaches you to break down any whole into its basic parts, to understand how those parts interact, and to see the unifying force that holds them together.

With this understanding, you can instantly discern fact from misalignment, recognize patterns, and navigate life with clarity. Confusion, noise, and contradiction no longer hold power over your thinking.

This book will guide you to perceive the balance in nature, society, thought, and in yourself. You do not need prior knowledge or authority, only curiosity and attention. This book is not just a philosophical model or spiritual reflection; it is a living pattern that points humanity toward its next great step, the awakening of a Level One Civilization.

If we learn to think, act, and relate through this balanced awareness, humanity may finally become what it was always meant to be, a species in harmony with itself and with its world. By the end, you will not simply know about the Trinity Balance in Reality; you will *see* it, and in seeing, you'll experience the quiet clarity that brings understanding, harmony, and awe.

Michael C Longo
Pine Mountain, Santa Cruz, California. Where the pattern first revealed itself.

Dedication

Author's Note

Table of Contents

Appendices, Acknowledgments, and Epilogue

Preface

While resting one afternoon in the Santa Cruz Mountains of California, a beam of light entered my awareness. Within that light was an image; a simple pattern showing the structure of reality as a Trinity. I was told that within this structure's understanding lay the answer to any problem humanity could face.

It was a message in words, eighty of them, but the experience was so overwhelming that I forgot them, trusting that if this answer to my prayers was true, I would discover it again, in time, through evidence I could find for myself. What I was left with was the promise that the light was true; it became a knowing, an understanding that unfolded in an instant and has continued to reveal itself ever since.

Through that moment, I came to learn that the pattern of the Trinity is not a religious symbol or belief system, but the actual framework by which reality organizes itself. Everything, from the turning of a galaxy to the beating of a heart, moves through the same simple law: balance through a three-part relationship.

This realization did not make me special; it made me responsible. Once you see how everything connects, you can no longer think that confusion, separation, and conflict are inevitable. Those thought patterns are symptoms of thinking outside the balance, of forgetting how to see the whole. And that whole, in any language, is known as God.

Many people experience God as a conscious being, a personal presence that speaks, guides, and judges. This book does not dispute that experience, nor does it claim one name or one form of God. Instead, it reveals the underlying pattern of creation as universal, interpreted differently by different peoples, yet reality itself does not differ. Reality can only express itself one way: through the balance of time, space, and motion. That is one Trinity. In Appendix A, you will find fifty-two Trinities; each showing how the same pattern repeats through every layer of existence.

Reality does not lie, but people can misrepresent God, and that misrepresentation has become the source of humanity's greatest confusion and division. Here, we will honor the pattern known as God, not as a

person or a name, but as the truth upon which reality itself is based. By doing so, we can begin to find our way toward becoming a level-one planet, a world where understanding replaces belief, cooperation replaces conflict, and love becomes the measure of intelligence. That is my aim.

To name the Source is less important than to feel the Source within you. The purpose of life is not to defend God, but to live in God; to dwell in that quiet current of love that holds all things together. Call it Spirit, call it the Tao, call it Allah, call it Consciousness, call it God; the names differ, but the law is the same. Love, born in effort, it is the glue of reality. When we understand love operating as a Trinity Balance, we no longer need to fight over whose God is true.

We see that all things are born from one Source, expressed through threefold balance, and completed in harmony. This is my answer to how humanity can become a level-one planet, where unity exists for all creatures, where understanding reality and loving each other is what we do above all else, and where cooperation to make life better for all becomes the rule, not the exception.

This book is my attempt to share how we might reassemble ourselves into a level-one planet, not through a belief system or religious following, but through an understanding of how reality itself works. When you learn to recognize the three basic parts in all whole things, confusion ends, and authentic love begins. What remains is awe; that sacred recognition that what is common to all is the love of life itself.

Michael C Longo
Santa Cruz, CA, 2025

INTRODUCTION

The Structure of Wholeness

Modern life has trained us to see the world in fragments. We study matter apart from mind, people apart from nature, and nations apart from their ideals. The result is brilliance without harmony, knowledge that multiplies but seldom unites.

Yet beneath all the noise and division lies a single reality that every part of existence obeys. It is the frequency of balance, the rhythm through which complementary forces create wholeness.

This frequency of life occurs through the **Trinity Balance in Reality**. It is not a belief system, but a first principle law of relationship: Two interacting complementary parts generate a third function that unites them, becoming a whole thing.

We see it in physics: **Time + Space → Motion = Reality.**
We see it in biology: **Energy + Form → Function = Life.**
We see it in ourselves: **Self-Image + Feeling → Consciousness = The mental state of being.**
And we see it in society: **Mother + Father → Child = Family/Society.**

Across every domain of existence, the same relationship repeats. The pattern never changes, only the scale and/or life form you are observing.

To understand this pattern is to restore clarity to thought and coherence to perception. Once you know whether something is a *whole thing* or a *part of a whole thing*, your mind stops spinning in contradiction. You begin to see how everything connects, how meaning flows naturally, and how truth feels, not abstract, but alive.

That moment when unity helps you transition from not understanding to knowing, from logic to intuition, is what this book calls **Awe**. Awe is not emotion alone; it is the resonance between perception and truth. It is the living point where logic, intuition, and unity agree.

This book is a map back to that state. The journey begins with my co-discovery of the three-part pattern itself, then follows its historical presence as a first-principle insight recognized by others across time. From there, it shows how to *see* the Trinity in anything, starting with the simple structure of a sentence, before moving on to how to dissolve confusion and finally into the lived mechanics of Awe.

This book isn't just about personal clarity; it's about *collective evolution.* When individuals align their thinking with the Trinity Balance in Reality, they become coherent with the same order that governs the universe. Imagine a civilization built on that shared coherence, one where competition gives way to cooperation, and confusion gives way to creative balance. That is how a **Level One Civilization** begins, not through technology alone, but through consciousness that mirrors the structure of reality itself.

My hope is not that you merely read about the **Trinity Balance in Reality**, but that you begin to *see through it,* to recognize its structure in every thought, every relationship, everything that you find in your life, so it is clear to you what you are seeing. Because once you see the pattern, confusion loses its grip. You begin to think as reality itself thinks; in balance, in rhythm, in wholeness. And that is when life starts to make sense again. That is where being in Awe becomes your natural state of being.

Part I

The Journey of Discovery

One

The Quest and the Question

There I was, standing on the landing, about to be taken to the airport and flown to Thailand. This was to be my first duty station as a newly trained Army medic. Joining the Army felt like a practical choice, two years of service, the GI Bill to cover some college costs, a chance to finally step out of the house I'd grown up in. But my friends had warned me, fresh out of high school, heading into the Army. They said I'd made a mistake. I didn't understand why.

Waiting on the platform with four others from my platoon, my mind buzzed with questions. What is truth? What is consciousness? How do I tap into my higher powers to become someone I could be proud of? All my life, I had been waiting for a chance to shine, to discover my real talents. Now, on the edge of this journey, the questions felt heavier than any duffel bag I carried.

A tall, confident soldier sat next to me. With hours to wait, we began talking. I rattled off my swirling thoughts about truth, and finally he said, "It's all about awareness."

"Awareness?" I asked.

"Yes," he said, leaning back. "It's being aware of what you are aware of. That's the highest truth."

That idea stuck with me as I flew to Bangkok and then 70 miles south to a tiny, perpetually-alert base. From there, we could barely see the B-52s taking off for missions over North Vietnam. It was 1972, and suddenly the world felt far bigger than I had imagined.

From my bunk, the smell of marijuana filled the air. Most of the guys were passing a joint around, part of a ritual I didn't yet understand. When it came to me, I hesitated, but figured, "Why not?" I'd smoked a little in high school. I took a couple of hits, then just half of the third. The room began to spin, nausea hit, and I vomited violently, blood mixed in, an unexpected consequence of the Thai weed laced with opium.

Lying there, drenched in sweat and trying to regain balance, I noticed my inner vision. I imagined seeing the rods and cones at the back of my eyes, and suddenly, there it was, a pattern, a perfect map of my visual field. The cells, the light, my consciousness, it all clicked. After enjoying this new found vision I drifted off to sleep, but only an hour later came something beyond imagination: my consciousness collapsed inward, launching me into a state that stretched my life back like a rubber band and then flung me forward at impossible speed into the present, and beyond.

Then, without warning or pre-thought, I found myself floating in outer space, a vast expanse of stars gave me a true sense of awe. My mind was perfectly still; I was observing with total wonder. As I drifted back into my body, I wondered what I had achieved. Was this the moment of full self-awareness? A glimpse into the ultimate truth?

Eventually, sleep claimed me. But the next day, wonder lingered. I told my mates I had gone to the fourth dimension. They nodded, accepting it as part of initiation. My awareness had expanded in ways I couldn't yet explain, fueled by that single act of curiosity, the desire to see the back of my own eye.

My duties were mundane: testing chlorine levels in water and delivering pills to prevent VD (STD's) among the women in the local town. But my mind had opened to something extraordinary, Surely the Army could use me in other ways. I had new skills after all. I was no longer a lowly medic; I was a star traveler.

And so began my journey; a question and a quest begun, and the first hint of a method that would guide me for decades to come, turning my attention and awareness inward. And I was ready to open and learn more about this wonderful world we live in.

Two

My Not-So-Great Escape

So, I suggested to my captain that I had 4th-dimensional skills that maybe the Army could use. Unbeknownst to me, the Army was looking for people who had such skills, so they took me up on my offer to do something else, but first I had to pass a test. I had to guard the monthly pay money for the locals with a gun. It was then that I somehow looked into the soul of what the Army really wanted and I saw some form of evil, and I in no way wanted to participate in that. I refused to use a gun to guard the pay money. I wanted out, to go to college and use my new skills to learn, to grow, and to discover what happened to me in the 4th-dimension excursion.

It was then that I remembered that the Army had a rule that if you served for 180 days or more and you could get out with an honorable discharge, if you were either crazy or hooked on heroin, you would have the GI bill money to attend college. So that was it, I decided to get out of the Army because I saw their soul and it was not a good place to be. So, given my Thai weed experience, I knew drugs were not an option so I chose being crazy as my means.

Two weeks of carefully staged odd behaviors, paranoid accusations, bizarre letters home, and purposeful silence were enough to convince the docs. I admitted myself, claiming I was paranoid and schizophrenic. It worked in Thailand, my duty station, but when I was sent back to the states, I was admitted to the psychiatric military hospital at the Presidio in San Francisco, CA. This was not going to plan. I had thought they would discharge me, but instead they interviewed me, agreed I had a problem and gave me a drug to take, daily, that clouded my thoughts, making me feel extremely tired. I could only sleep for nearly two weeks and had a hard time walking.

Then, one day on the beach, clarity returned. I could run, I assembled a 500-piece puzzle, proof that my mind worked. By day 29, I found myself mediating between a patient and a doctor, explaining what neither could understand about the others perspective. A nurse interrupted, walking me down the hall saying: I could go home now. Relief washed over me, but questions lingered, had they ever really understood me? Twenty-nine days in that place, why did no one say a single thing to me as to what they thought was wrong or how to fix my mind? I felt so guilty and alone. If their evaluation were correct, you'd think they would have said some words of therapy to me, right?

Nine months later, I finally arranged my early out that I qualified for serving near Viet Nam. My captain said, ok, you can have your early out, just go home and wait till I contact you. Living near Pebble Beach, I bought a convertible, an Austin-Healey and reveled in freedom for three months while waiting for my discharge papers. A mate spotted me while driving my Healey sometime after, shouting, "The whole base is looking for you!" I was surprised to hear this since I was following the last order I got. I shaved and cut my hair and reported in as instructed the next morning.

Discharge paperwork forced a choice: mental disability, 100% or 10%. Neither was right, I had faked it all, but I was not given the choice of receiving nothing, I had to choose. It was one of the hardest decisions I have ever made, to carry a label like that is scarring for the rest of my life I thought, so I took the lesser of two evils, a 10% rating. Shame lingered, but freedom felt sweeter than I could have imagined.

Back in civilian life, I dove into states of consciousness, reading everything I could. One day, meditating on top of a washing machine at a laundromat, a stranger asked about my practice. He told me about a 16-year-old master in India teaching four paths to enlightenment. A thousand-mile trip later, I learned the methods, seven years to enlightenment through any one path. I would choose inner light since the Bible in John 13.4 quoting Jesus said that "when the eye is single the whole body is filled with light."

Then came a pivotal invitation: a family living "on the land" in the Santa Cruz mountains, Last Chance Road, invited me to stay with them for a time. It was great but I needed my own place to do my inner work, so I found a small cabin own by Grace Slick's father, a patent attorney in San Francisco. I asked him if I could caretake his property and he said yes.

It was lonely at first, but I had all I needed. After a year of meditation four hours a day, I found myself asking the universe for the one single answer that could solve any problem in the world. Then there was a flash of light one evening, a download of the trinity image. A voice said about 80 words to me explaining that this is the one single answer that can solve any problem in the world but I was so overwhelmed with the rush of visuals and information I could not retain any of the words, only the impression that the triangle of three parts is the answer. The visual impression burned into my awareness. I rationalized that not paying attention to the 80 words was because if this is true then I would find the proof elsewhere and then be able to restate those words a later date. I had no idea it would take me this long to re-articulate the meaning of those words.

Three

The Three-Part Pattern is Confirmed

In the months of quiet meditation and observations that followed, I began noticing the three-part pattern everywhere. The first time I actually saw the wholeness and benefit of the Trinity was when I observed it in conversations. The two opposites were the two people talking. But, what was actually going on? It was what occurred between them, the quality of the energy they created, the third part is the most important element. The conversation wasn't about either of them, it was about what they produced. That product was the frequency of their combined conversation. I was awestruck. Life had a rhythm I hadn't fully appreciated before, and the Trinity Balance offered a lens to see it clearly. The tension between opposites, contraction and expansion, giving and receiving, observation and reflection, wasn't just a theory anymore. It was alive, pulsing in everything around me, waiting for someone to pay attention.

Then I found the mother lode of confirmation. I had been asking the Universe for confirmation and one day while in the library I found that proof in the form of Stewart's book, The Unobstructed Universe. Let me explain how it came into being. In the quiet years of the early 20th century, long before the digital age and its instant noise, there were three souls, Betty, Joan, and her husband, Stewart Edward White, who had all asked the same question that has haunted seekers across time: *What is the true nature of reality?*

I first need to set the stage. Betty, a very experienced medium (someone who communicates with those without a human body) had passed beyond the veil, into what she called "the Unobstructed Universe." Yet through the dedicated sensitivity of her friend Joan, she could speak again, slowly, deliberately, as if translating from a different density of being. Stewart, patient and exacting, acted as the scribe taking notes as he witnessed this impossible collaboration between two worlds.

It was not easy work. The effort to bring such information through was immense. Betty herself explained that communication between planes was like *trying to shout across a roaring waterfall*, words dissolved, meanings twisted, and energy drained. But she persisted, driven by the certainty that humanity was ready for one revelation above all others, the only teaching she said she was permitted to bring back:

"All reality, yours and ours, is built upon three essences. Receptivity. Conductivity. Frequency. These are the true constituents of reality."

To Stewart, the idea seemed impossibly abstract at first. What could it mean that *time itself* was Receptivity, that *space* was Conductivity, and that *motion*, the living pulse of all things, was Frequency? But as months turned to years, he began to see how these words mapped perfectly onto both the seen and unseen universes.

Betty explained that in our "obstructed" reality, everything manifests as Time, Space, and Motion, the measurable frameworks through which existence expresses itself. Yet behind these are their living counterparts, the *essences* of being that give rise to the forms we experience.

Receptivity, she said, is the **essence of Time**, the inward drawing, the stillness that allows perception itself to occur.
Conductivity is the **essence of Space**, the medium through which awareness connects and relationships are formed.
And Frequency is the **essence of Motion**, the vibration, the creative oscillation that gives all things life.

To understand these was not a matter of theory. It required what Betty called *alignment of consciousness*, a delicate balance between inner silence, mental clarity, and emotional harmony. Stewart often described how exhausted Joan would be after a session, how Betty would pause for long minutes before speaking again, as though adjusting her entire being to maintain a signal clear enough for truth to pass uncorrupted.

Each domain revealed itself slowly, through lived experience rather than through concept. Receptivity was learned in stillness and surrender; Conductivity, in effort and discipline; Frequency, in the joy and awe that arose when the other two harmonized.

That is why their work endures, not as a set of claims, but as a living model of what it means to listen deeply across the boundaries of reality. Betty, through sheer devotion, transformed metaphysics into method. She showed that consciousness itself operates by the same trinity that underlies light, sound, and form.

And this, too, is where my journey continues hers. When I received my instructional vision, the triangle of balance, the revelation that every problem in the world could be solved by seeing how its three opposing yet complementary forces form a single unity in anything that I study, I was stepping directly into the same stream Betty was living. The message had not changed; it had only evolved its messenger.

What Betty was permitted to reveal, I was destined to extend: that Receptivity, Conductivity, and Frequency are not only the structure of the universe, but also the **pathway to awe**, the living proof that consciousness is woven through all things.

Every great thinker has glimpsed this truth in their own language.
Plato's *three-part soul.*
Newton's *three laws of motion.*
Tesla's obsession with *3, 6, and 9.*
Freud's *id, ego, and superego.*
And now, thanks to Betty, Joan, and Stewart, and now through me, the ancient Trinity is reborn with new clarity.

It is time, again, to understand that everything, from the smallest atom to the largest idea, expresses itself through the tension and unity of three interacting domains. And when we learn to hold that balance within ourselves, to become receptive, conductive, and vibrantly alive in frequency, we touch the same unobstructed truth that Betty once reached across worlds to share.

Betty once said that comprehension alone was never the goal; *alignment* was. Knowledge without experience is like reading a map and never walking the trail. To know the three domains is one thing; to live them is another.

And so, as we step from the legacy of Betty and Steven into our own moment in history, the question becomes personal: how do we *feel* the structure of truth within us? How do we turn Receptivity, Conductivity, and Frequency from ideas into a rhythm that reshapes our lives?

The answer is to experience awe. Awe is the living intersection of all three, the doorway through which understanding becomes transformation. It is where awareness receives, connects, and vibrates all at once, restoring balance to both the inner and outer worlds.

The Key Revelation

The key takeaway from these three chapters is this: what was given to Betty from the dimension she now inhabits, and what was given to me from the dimension beyond this physical world, are in truth one and the same message, delivered through different doors of perception, yet pointing to the same infinite truth.

Betty was told that the Trinity was the only thing she was allowed to reveal, the one truth that stood above all others. And I, too, was given only one answer, and it was the same: *the Trinity*. I was told it was the key that could solve any problem in our world.

Two revelations, each received in isolation, yet unified by their message, the same truth emerging through two witnesses of different realms. This is not coincidence; it is confirmation. It means the pattern is not human invention, but a law of reality itself; the structure of consciousness that unites all things.

These parallel revelations form a bridge between worlds, a living demonstration that balance is the nature of truth, and that understanding this balance is how humanity may begin to rise toward its next stage of development: the first level of civilization living in conscious harmony with the creative forces of existence.

Let this awareness settle deep within you. Feel its weight, its promise, its urgency. What was whispered through Betty and shown to me is not a theory, it is a key. And with it, humanity can unlock its next and greatest chapter.

Part 2

The History of Trinities and Their Domains

Four

Historical Trinities of Wholeness

Every great thinker has glimpsed the truth of the three-part pattern. They saw it not as dogma, but as the natural order of balance. Each spoke it in their own language, in the symbols of their own time, yet they all pointed toward the same center, the way that unity expresses itself as three.

Plato was among the first to name it clearly. In his vision of the soul, he saw three governing parts: reason, spirit, and appetite. When reason leads, spirit supports, and appetite follows in harmony; the soul is what they become, the whole. But when any one part dominates, confusion and imbalance arise. Plato's triad was not only a map of the human psyche; it was also a model for society itself. His ideal republic mirrored the soul: the philosophers as reason, the guardians as spirit, the producers as appetite. Order was balance. He promoted another Trinity, that of thought. He called them the ultimate ideals; perfect essences of reality: goodness, truth, and beauty.

Centuries later, **Plotinus**, the mystic philosopher, described reality itself as a Trinity: *The One, the Intellect (*or Nous)*, and the Soul* (or

Psyche). The One was the source of all, Intellect the pattern of all, and Soul the living motion that carried it into the world. His Trinity was not an invention but an observation: all things flow from unity into duality and finally into living threefold-ness; wholeness expressing itself through relationship.

In the world of science, **Isaac Newton** discovered that motion itself obeys a Trinity. His three laws describe how every movement contains a balance between cause, continuation, and response: an elegant symmetry that holds the universe steady in change. Even here, in the cold precision of mathematics, the threefold law stands.

Nikola Tesla, the visionary inventor, was haunted by the power of numbers: especially 3, 6, and 9. He said that understanding these numbers would reveal the secrets of the universe. Tesla sensed, perhaps intuitively, that creation itself unfolds in threes: charge, field, and motion - past, present, and future - birth, life, and transformation. I like his use of three the most of all because his inventions are applied examples of the balance of the three forces in nature.

Sigmund Freud brought the same pattern into the modern mind. His model of the psyche - *id, ego, superego* - is another reflection of the same ancient form. The instinctive self, the balancing self, and the moral self: three voices negotiating wholeness. When in balance, the person is integrated. When one dominates, suffering follows.

Across time, cultures, and disciplines, the triad reappears. The **Hindu Trimurti**, Brahma the creator; Vishnu the preserver; Shiva the transformer. The **Christian Trinity**, The Father, Son, and the Holy Spirit. In the natural world, we see it everywhere: solid, liquid, and gas - seed, growth, and fruition - morning, noon, and night.

Hegel, the German philosopher, turned the threefold pattern into a logic of becoming. His *dialectic*: thesis, antithesis, synthesis, describes how every idea evolves through tension and resolution. Two opposing yet complementary forces meet, struggle, and finally unite in a higher form that includes them both in synthesis. This process repeats endlessly, driving not only thought but history itself toward greater integration. To Hegel, the triad was not merely structure; it was movement, the pulse of reality seeking self-understanding.

In the East, the pattern was seen in softer light but equal depth. **Taoist philosophy** speaks of *Heaven, Earth, and Human* as the three harmonizing forces of existence. Heaven is what we are receptive to, Earth is what conducts and holds life, and Humanity serves as the bridge that joins them. When the three are aligned, life flows effortlessly: the Tao made visible through balance.

The **ancient Egyptians** saw their world through trinities: Osiris, Isis, and Horus: life, love, and renewal, the eternal cycle of creation continuing through relationship.

In the **Christian era**, the Trinity took sacred form: **Son, Father, and Holy Spirit**, a sequence that reflects the same three-domain structure found throughout reality. The Son corresponds to the **time domain**, the dynamic flow through which creation unfolds. The Father corresponds to the **space domain**, the stable knowledge or form that holds what is created. And the Holy Spirit expresses the **unifying domain**, the integrative function that arises from the relationship of the first two (see Appendix A). Theologians wrestled for centuries with how one God could be three persons, yet their debates echoed the same mystery that Plato and Plotinus grappled with: that unity and multiplicity are not contradictions, but complementary forces in the ongoing dance of being.

Modern science, with its material rigor, also discovered the triad. The atom itself**: Proton, Electron, and Neutron**, embodies the stable tension of three. Energy, Matter, and Information: the new trinity of physics, define every process from the birth of stars to the thoughts in our brains.

Language itself reflects this triadic logic: past, future, present; beginning, end, middle; verb, subject, and sentence. To think in threes in the correct order of creation is to think in wholeness. Recognizing the emergent third force allows us to understand how seemingly opposing elements can converge into a coherent whole.

The Trinity Balance, therefore, is not a theory among theories. It is a lens through which the universal mechanics of reality, in thought, matter, emotion, and society, can be observed, understood, and applied.

With this work, the Trinity as seen in the lands of ancient Europe to the laboratories of modern science, from the myths of Gods to the structure of the mind, can continue through your understanding and applications. Every age, in its own tongue, rediscovers the same geometry of completeness. When we look closely, we see that these trinities are not coincidences or poetic flourishes. They are the fingerprints of life's balance itself, the recurring rhythm through which reality holds itself together.

This is the rhythm is the soul, the atom, the family, the nation, and the cosmos we all seek. It is the pattern that both ancient seers and modern scientists have traced in their search for order amid chaos.

And now, from those historical figures, the lives of Betty, Joan, and Stewart, and through my own discovery of this pattern, the ancient Trinity is reborn with new clarity. Not as a borrowed symbol, but as a living truth, something that can be felt, applied, and lived. The Trinity Balance is not a theory; it is the very grammar of reality itself, and when we learn its language, confusion becomes understanding, and chaos becomes Awe.

Five

The Domains of the Three Parts

You've seen how great thinkers throughout history recognized trinities within the wholeness of our world. Now, let's look at these three parts as **Trinity domains**, the foundational ways to view reality itself. Understanding these domains will bring clarity to your thinking and help you make sense of anything, from atomic physics to everyday life.

Each Trinity domain represents one of the three complementary parts that make up every complete whole. You can think of these domains as the grammar of the universe: the underlying order that defines how every whole thing is structured. This is how the world maintains coherence and meaning.

In this book, Trinity Domain, is a term used to describe one of the three complementary parts of a larger whole. Each domain contains a set of descriptive elements that together express the full character of that Trinity domain. To use an analogy, think of your home as one of the Trinity Domains. Just as every object in it, the furniture, books, tools, and personal items, when you think of them, notice how they instantly remind you of your home. Those are examples of how different terms are used to describe each of the Trinity domains or the pattern of three in our reality by thinkers and scientists in our historical record.

Below are the three Trinity Domains and the whole they form: two complementary domains and one unifying domain. Read the table both vertically and horizontally to see how nature's pattern unfolds as descriptive elements. Notice their order, that is essential to working effectively with trinities in reality. The first three are the usual inward-to-outward way to read a trinity. The last two are out of order for reasons we will cover in a moment. As you practice seeing this pattern, you will find your Awe moments naturally arising, filling your awareness with their light and clarity.

Domain 1	+	**Domain 2**	→	**Domain 3**	=	**Whole Thing**
Time	+	Space	→	Motion	=	Basic features of the universe
Receptivity	+	Conductivity	→	Frequency	=	Basic forces in the universe
Center	+	Circumference	→	Space	=	Shapes
Son	+	Father	→	Holy Spirit	=	Catholic Understanding of God
Error	+	Trial	→	Learning	=	Knowledge

Notice how the last two trinities are out of the usual order we hear them. These two trinities, when read in the correct sequence, from inner to outer, from small to large, and from time to space, reflect how people understand creation. Creation always moves from the center outward, producing a frequency, a rhythm of being.

The Church, however, traditionally states the Father first, then the Son, then the Holy Spirit. This top-down presentation reflects how God taught people, but it is not aligned with the natural order of creation. Understanding what exists more in time first then what exists more in space helps us see this order of creation clearly. Always reading from inner to outer prevents confusion.

When we observe the parts of a whole in this order, we organize our mind correctly. First think of the time-related domain, then the space-related domain. When these two complementary parts are combined a frequency of their whole being is created. Once you understand this sequence, you can recognize whether something is being presented from a top-down perspective, as God would, or as a human presents it. Many people, unaware of the correct sequence, repeat what they were taught without considering how the mind perceives this natural order. Recognizing this difference is how we develop our "mind's eye of wisdom."

In **Appendix A**, you'll find fifty-two trinities listed primarily in the human, inner-to-outer direction. The bottom-up method of saying this trinity, Son, Father, and Holy Spirit, sounds so odd due to the way we have been hearing all our lives, but that is the correct order, inner to outer, then the unifier, the holy spirit, this is the order of creation. Being aware of both sequences allows you to view the parts of reality in the all-important order in which creation occurs.

Reading reality in these two ways forms can be visualized as a six-pointed star: two overlapping trinities, one pointing upward (the human perspective) and the other pointing downward (God's perspective). This distinction is crucial for understanding and applying the threefold nature of reality.

So, whenever you examine reality by splitting it into its three parts, a first-principle approach, be conscious of your perspective. Are you looking from the human viewpoint, gazing up at creation, or from the perspective of top down, first seeing the whole thing, the space it occupies, then the time domain outlining how long it took to be created? Knowing the direction is essential for clarity and alignment with reality's natural order and a good habit to develop.

Star of David

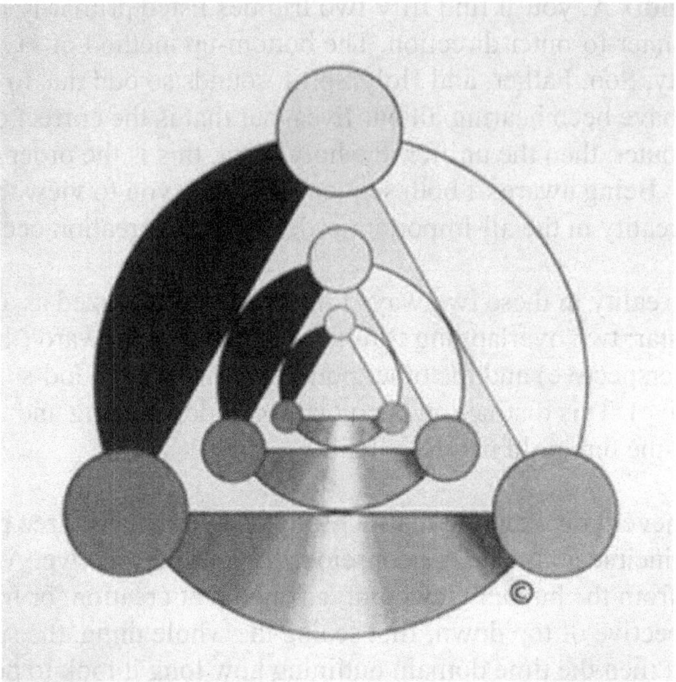

Interpretation of the Book Cover Image

The three circles of different sizes represent the different levels of reality. The three smaller trinity circles, in the primary color's red, blue, and yellow, represent the three basic elements found in all whole things in life. Black and white represent the absolute opposites or complements of reality from which all things or the rainbow emerges. The rainbow and yellow symbolizes the unity in all things. The triangles represent that everything in our world arises from three parts. The entire image bursts outward in yellow, showing that light itself has at its core a Trinity Balance, and that all things originate from light.

Deep thinkers from all over the world have captured the truth in the reality of the trinity balance. Here are a few more images to enjoy.

Tibetan Mandala

Yin-Yang Symbol, typically done in black and white with no star burst, but I like this one because it adds more content.

The Celtic Trinity Knot or Triquetra is the knot means the three parts are interconnected and cannot be broken in any whole thing.

Showing the three parts of reality is easy as an image; harder is expressing them in conceptual words, and hardest of all is identifying the three parts in a whole thing as you analyze it. That, however, is exactly what we are doing here, what Betty accomplished for us in the 1930s, and what many historical religious, scientific, and philosophical thinkers have pursued throughout history. This book captures the best of that knowledge for your first-principles understanding, in the hope that you will use it wisely, aligning yourself with life and allowing your love energy to flow in harmony with it. To go beyond this foundational understanding may lead to dead ends for you as one researcher has discovered: See Appendix E: Why the Trinity Balance is Enough.

Part III

The Trinity Balance as a First Principle Means of Understanding

Every journey begins with curiosity, which shifts into learning, then knowing, and finally into capability.

Up to this point, you have explored the deep history of trinities, how every culture, every age, and every wisdom tradition has reached toward this same structure of wholeness. You have also seen that each part of a whole corresponds to a distinct Trinity domain, expressed as descriptive elements shaped by human interpretation and perception.

Part III now moves from background to practical ability. Here, the Trinity Balance becomes a tool you can use: a first-principle method for seeing wholeness in any thought, any situation, any conversation, even for clearing your own confusion. Once you learn to read a Trinity Balance as it appears in a sentence, your inner landscape becomes clearer, providing a foundation for action. And once you can read it, you can rebalance it when necessary.

This part is about gaining fluency. It's about discovering that understanding is not an accident, but the result of effort applied to structure. And when you see that structure, confusion ends, not by force, but through clarity.

Six

Reading a Trinity Balance as a Sentence

The trinities sentences that follow will help you understand how to think of each whole thing and its parts. This is a First Principle method for organizing your mind, one that leads to clarity of thought and can end the confusion or "stuck" patterns of thinking you may have. As you read these sentences and consult the Master Trinity List in Appendix A, you will begin to see how things align as reality truly is, in nature, society, and the workings of your own biology.

As we learned in Chapter Five people naturally read a trinity from the center outward, while the divine, or "God" way, is from the outside inward, as seen in the reading of the Father (outer), the Son (inner), and the Holy Spirit. Notice how awkward it feels to reverse this order, reading the Son (inner), the Father (outer), and the Holy Spirit. This awkwardness arises because it conflicts with how you have traditionally encountered it, if at all. The first step, then, is to order the trinity parts correctly in your sentences, from center to outward.

Most trinities in this book follow this center-outward perspective, and I encourage you to do the same. When you understand the essence of a Trinity balance, this distinction becomes clear, and you will see why reading them in the order of creation produces the most natural flow of understanding. Your mind will become clear as a bell as you explore this First Principle method of comprehending how reality is organized.

Each trinity in the following sentences are noted in bold type: the first two words in bold are the **complementary opposing** parts, the first two domains, and the third bold typed word is their **unifier.** The fourth bold typed word is the **whole thing** they equal. They are numbered to correspond with the Master Trinity List in Appendix A. Many trinities made little sense to me until I could read them as full sentences. As you study the Master Trinity List, try constructing a few of your own trinity sentences; this is the best way to internalize the Trinity Domains and see the patterns of the world. Focus especially on the main Trinity domains of time, space, and motion, as well as receptivity, conductivity, and frequency, to guide your understanding.

To complement the horizontal reading of a single trinity describing a whole thing, try reading them vertically, one Trinity Domain at a time. Doing so will quickly reveal how descriptive terms are inherently limited in scope; they are tightly bound to the whole thing they describe, which is compelling.

The Emergent, Dependent Force

A special note goes to the unifying factor in every trinity. Notice how it is always invisible, an emergent force, neither good nor bad, simply arising from the interaction of the complementary forces themselves. It is the silent axis that allows the other two parts to exist and interact. Its presence is rarely seen, yet it is essential. This often explains much of the confusion and imbalance in our world, because it is frequently overlooked, forgotten, or ignored.

Learn to recognize and work with this force as a bridge between complementary forces in your life. It is not neutral in the sense of being independent; rather, it is entirely dependent on the two active forces. Yet it is completely natural, arising spontaneously whenever the complementary forces interact.

1. Things exist in **time**, occupy **space**, and set anything they affect *in* **motion**, becoming **observable participants in the reality** in which the universe exists.
2. Things are **receptive** in time, because it takes time to receive anything. Space **conducts** the energy of time: you cannot have one without the other. Together they combine as **frequency**, forming the **fundamental forces** of our universe.

5. When things are in **contraction**, they are compelled to **expand** again, guided by the larger **self-balancing forces** in which they exist: **the Living Universe.**

7. The **proton** rests in the nucleus of the whole thing called an atom, while the **electron** forms the field that occupies up space. Together, they combine to create the **neutron**; evident when the neutron is split, revealing a proton and an electron within it. All three combine to make an **Atom**.

(A nearly massless *electron antineutrino* is also created in this process, conserving energy, momentum, and lepton particle. For our purposes, however, this fourth particle can be set aside, since all other trinities do not include a lingering part that falls outside the First Principle focus of clarity.)

9. Every whole thing has a **center**, where it contracts in time, and an **outer circumference**, where it expands into space. The **region** between them defines the space the **whole thing** occupies.

12. The **mass** of something **traveling at the speed of light** due to a form of **propulsion** is a form of **energy**. (Notice how this trinity had to be spoken with the space element first, then the time element as speed finds propulsion which has unified it that then can be measured as an energy value.)

14. The contractive force of **magnetism**, when paired with a complementary conductive element such as **steel** (which provides additional electrons to their unity), produces a dynamic interaction. When **manmade motion** causes them to pass near each other **Electricity and Light** are generated for use in our world.

15. The contractive **Yin** force of the universe is balanced with the expansive **Yang** force, always forms a unity. When some form of **energy** is applied, such as eating food or exercising, these forces create their own unique **state of health within you.**

18. **Concentration** unfolds in time, and **relaxation** unfolds in space, both guided by the energy of your **intention,** which in turn defines the **quality of your cognitive function** at that moment.

21. **Entropy** is a contractive principle, expressing that everything decays over time. Yet we observe life that grows despite the negative aspects of entropy, defines **positive entropy** in **living beings** of all kinds. Both of these trends in nature are recognized as integral parts of the **cycles of life.**

22. Scientists have found that people have a logical **left-brain** hemisphere and an intuitive, focused **right-brain** hemisphere. Together, they create **cognition**, the process of thinking, from which **consciousness** arises.

24. To be **unconscious** is to exist in your most contractive state. Add **consciousness,** and you gain **perception:** all three together form your **awareness** of your surroundings.

28. Using your **inner map** of yourself, combined with the **self-images** you have concluded that you are, constitutes your **self-knowledge** in that moment. Add more and more of these moments together and you cultivate **awakened self-awareness.**

30. When you have a deep **question,** you pose it to the universe. Over time, your question attracts all the **knowledge** related to it, even while you are **wondering** what the answer is. Eventually, the **answer** arrives, and you are filled with Awe upon learning it.

32. A **mother** exists in time as she nurtures and builds the generational family. A **Father** creates the space necessary for that structure and the **child** unifies their shared purpose. Together, this is what is needed to create a **family**.

35. When you see something that **disgusts** you, it triggers a deep inner feeling, which you then express outwardly as **hatred**. Both of these reactions stem from a **judgement** you have made, what we can call **negative emotions**.

36. The United States government is designed for **Democrats** and **Republicans** to **debate** each other's perspective in order to reach the best solution to issues that best serves the peoples needs. This is how **good governance works**. (That's how government is intended to work. I discuss how actual good governance can be achieved by making the Preamble to the Constitution binding Law on my YouTube Channel, The Way of Awe.)

39, **Verbs** are actions words that unfold over time and require a **subject** to act. Both are necessary to form a **sentence,** which expresses a **complete thought**.

41. The **Son** over time learns from the **Father** who occupies all space. They communicate through the **Holy Spirit,** reflecting how the Catholic Church understands our **relationship with God.**

44. **Death** happens in an instant, while **Life** unfolds in space, and **Rebirth** connects the two complementary opposing forces. These are the **cycles of all living things**.

49. One's **self-image**, combined with your **talents,** you **express,** defines the quality of your being. This is the most common way to describe a **human being**.

52. **Heaven** exists in a different dimension of time that we cannot access in the here and now. On **Earth,** however, **humanity** can create such a place if we choose. We like to call this place **Paradise** and dream of it.

Reading Trinities in Practice for Understanding and Awe

When studying these trinities, do not rush to memorize them. Instead, feel how each relationship breathes; how, for example, #5: *contraction, expansion, and the force that complements them* create a larger whole thing. Over time, you will sense this rhythm in every situation, every whole thing, and every living process in reality.

To read a trinity is first to analyze its parts, and then to recognize the true nature of the whole they form. You are dissecting reality as a First Principle activity so you can witness how it is actually organized, not how you wish it were, and not how others insist it must be. Every whole thing, whether an atom, a thought, a family, or a nation, moves through and expresses its own Trinity Balance of parts.

When you learn to see in this way, confusion dissolves. Complementary forces cease to battle; they begin to harmonize. The emergent force between them arises naturally from their interaction, entirely dependent on the two foundational parts that appear opposite in nature yet are, in truth, deeply complementary. As you learn to express a Trinity Balance as three words or as a sentence, your mind finds its natural order, and your heart finds its natural peace.

Every moment in life offers this structure. Observing it quietly lets the Trinity Balance guide action without confusion. You begin to see tension as opportunity, complementary forces as partners, and emergent understanding as evidence that balance is possible.

Imagine walking home, watching wind ripple through trees, hearing laughter in the street, feeling the gentle push and pull of your thoughts. Each moment becomes a chance to recognize, participate in, and experience the living rhythm of the Trinity Balance, the breath of life.

When this perspective becomes natural, awe arises spontaneously. Not because something extraordinary happens, but because ordinary moments reveal reality's true depth. Life itself becomes a teacher: each tension a lesson, each interaction a mirror of the underlying pattern of wholeness.

Finally, once you feel the pulse of the Trinity Balance in every whole thing, the next step is learning to recognize when forces are misaligned, and how to restore clarity. This is the important subject of the next chapter: **Using the Trinity Balance to End Confusion.**

Seven

Using the Trinity Balance to End Confusion

Confusion is everywhere, in the world and sometimes in our own mind. Leaders act according to what they believe is right for their country, which often conflicts with what we feel in our hert is right. And at times we, too, are unsure how to respond or understand what we are experiencing. Both forms of confusion share a single cause: a failure to see the whole situation while seeking a benefit for one and all. Since we cannot know the hearts or minds of leaders, this chapter focuses on what we can do as individuals to end confusion. The first step is understanding the laws that govern the Trinity Balance. Knowing these parts allows you to see the whole thing, or subject, which is the path to understanding and enlightenment.

The Laws of the Trinity Balance in Reality

Law 1 Composition of Reality's Parts

Every whole thing consists of three parts. Two are complementary counterparts, whose interaction produces a third component, a frequency arising from their relation. The structure and behavior of any whole thing are determined by the alignment and interaction of these three parts.

Law 2 Emergence of the Unifying Part

The third component is an emergent frequency generated by the unique qualities of the two complementary parts. It unifies the first two into a single definable whole. This component is energetic, not material; it can be detected only through its effects, not by direct observation.

Law 3 Order of Creation

A trinity forms from the center outward. This order defines the natural direction of life and creation. Reversing this order is possible, but it produces disorder and confusion.

How to Use These Three Laws

To understand any whole thing, first break it down into its three parts. Appendix A provides examples. This first-principles approach can reveal the structure of life and the universe, and even produce moments of awe. Focus first on identifying the three parts in the correct order; doing so creates clarity, while reversing them produces confusion.

Confusion often occurs when parts from different whole things are mistakenly treated as if they belong to the same whole. For example:

- **Trinity #17:** feeling truth + understanding truth → empathy = love
- **Trinity #19:** pessimism + optimism → emergent emotion = emotional elasticity

Treating parts from #19 as if they belong to #17 leads to misunderstanding. The complementary parts do not align, and the emergent whole differs. In #17, feeling truth combines with understanding that truth, which conducts the feeling of knowing. Their interaction produces empathy; the whole is love. In #19, pessimism and optimism combine with the emergent force of emotion to create a person with emotional elasticity, not love.

Actualizing the Three Laws

To experience the laws in practice:

- Read the Master Trinity List (Appendix A) from **left to right**, identifying the three distinct parts that form a whole thing.

- Read the Master Trinity List **right to left**. Seeing the whole first can help you identify the right parts of the subject you are exploring.
- Read the list **vertically**, noting that most parts belong only to the numbered trinity pairing.
- Observe carefully: first-domain parts may superficially resemble parts in another trinity, but true alignment occurs only when complementary parts interact as intended.

This exercise trains your perception to recognize alignment versus misalignment in any whole thing.

Three Ways to End Confusion

1. **Identify the right trinity of parts**
   Find the two complementary forces that belong together, generating the emergent frequency that defines the whole thing.
2. **Put them in the correct order**
   The center, complementary, and unifying parts must align properly. Follow the sequence of creation as Domains 1, 2, and 3 indicate.
3. **Recognize the emergent third**
   Confusion arises whenever one or more parts of the trinity are overlooked, whether only Domain 1, only Domains 1 and 2, or any incomplete combination. Learn to appreciate how the third part emerges naturally from the interaction of the complementary forces. It depends entirely on the first two and cannot exist outside their union.

Seeing all three parts in any whole thing is First Principle understanding. Breaking things down this way reveals how the universe orders itself and allows clarity and peace to emerge naturally.

A Physical Example: Magnetism

At first glance, magnetism seems to have two opposing poles:

positive and negative. Many people think of the poles as independent, endlessly chasing the tension between them. Yet both poles are part of a single complementary force: contraction.

The true interacting force with magnetism is a conductive medium, such as steel. When magnetism interacts with steel and motion is applied, electricity and light are created. Confusion arises if we see only the two poles of magnetism and not the conductive domain or the man-made force that caused them to move. Recognizing how the two interact tells us what it takes to create the motion needed to produce a desired amount of electricity or light.

Trinity expression: % of magnetism + % of conductive medium → applied motion between the two = % of electricity/light.

An Emotional Example: Feeling and Thought

People often see feeling and thought as opposites: heart versus head, emotion versus logic. Culture teaches us to choose one or the other.

In truth, they are complementary forces:

- **Feeling** is receptive, absorbing and generating a sense of time.
- **Thought** is conductive, structuring and interpreting patterns into knowing.

Together, their interaction produces understanding, the emergent effect. Confusion arises when one dominates, or they are pitted against each other. When they flow together, heart and mind act as one, and contradiction transforms into coherence.

A Social Example: Authority and Freedom

Societies swing between authority and freedom. Each claim to remedy the other: more control prevents chaos, more freedom escapes control.

Authority contracts, it provides structure and limits; f**reedom expands**, the container of individual expression. Alone, neither achieves balance. Only through their **interaction**, guided by integrity, does social coherence appear. This can be measured by the **quality of life in a society**.

The U.S. government was originally designed to function this way: complementary forces debating to find the best path in service of the people. Evaluating how well each branch fulfills its role is essential to maintain balance; without it, chaos can ensue. The rule is simple: extremes cannot rule the day. Balance is our best guide, just as the founders intended.

Trinity expression: Authority + Freedom → Honest Debate = A Governed Society

Conclusion

Once the two complementary forces are identified and the emergent third recognized, clarity returns. The three-part pattern is consistent across all domains: time, space, and motion.

Confusion is simply misalignment or a missing emergent result. Once seen, the mind quiets. Effort gives way to recognition. The Trinity Balance becomes a living compass, pointing toward wholeness.

Each reconciliation of complementary parts produces a subtle realization—a felt recognition of unity. This is the beginning of Awe. Creation flows correctly from center outward, from feeling to form, from freedom to structure, from heart to mind. Moving against this current creates resistance; moving with it aligns you with life itself.

In the chapters ahead, we will explore how this same order of creation operates through perception, awareness, and the awakening of **Wholiocognition**, the practice of immersing yourself so fully in your own energy that ordinary experiences become moments of awe.

Part IV

The Way to Understanding, Life, and Awe

Part IV is where the Trinity Balance comes fully alive. Up to now, you have learned the structure of complementary forces, the domains they inhabit, the role of tension in creating a unifying element, and how every whole thing expresses these three parts. You have also seen how this knowledge can dissolve confusion.

Now we move from theory into practice. You will learn the method, the Trinity Balance Formula, and how it can be applied to any idea, event, or challenge. You will explore real-life stories that show how others navigate conflict, relationships, and complex decisions using this approach. Finally, you will discover how understanding becomes lived awareness, how the rhythm of the formula becomes an inner capability, and how this unfolds into moments of Awe. Understanding → Life → Awe. This is the journey of Part IV.

Eight

The Trinity Balance Formula

Every insight in the previous chapters leads to this one.
You now know that every whole thing is made of three parts, that complementary forces create an emergent unity, and that imbalance, whether within a person, a relationship, a system, or a society, always signals that one part is overpowering the others.

This chapter introduces the structure that makes these ideas practical: **The Trinity Balance Formula, a simple, mathematical way to see how any situation works, why it feels the way it does, and what must change for clarity or harmony to return.**

The formula gives shape to something the mind senses intuitively: that all problems are imbalances, and all solutions are rebalancing. By identifying the two complementary forces at play and the unity they generate, you can adjust the proportions, correct distortions, and create a whole outcome that aligns with your intention.

This method is not theoretical.
It is a tool you can use to analyze any experience, decision, conflict, emotional pattern, or challenge. Whether you are working with inner life or external circumstances, the Trinity Balance Formula gives you a clean, repeatable way to understand what is happening and what needs to shift.

With it, understanding becomes structured rather than accidental, intentional rather than mysterious.
And clarity becomes something you can build.

The Common Feature of Every Problem

Every problem, no matter the scale, has the same structure: something is out of balance.
This applies to thoughts, emotions, decisions, relationships, organizations, and nations. When a system stops working, one part has taken on too much influence, another part is under-expressed, or the ratio between them no longer fits the goal.

Rebalancing sounds simple, but it rarely feels simple.
Most people try to fix the *symptom* of imbalance instead of identifying the two forces that created it. But once you identify those forces, rebalancing becomes straightforward. Appendix A lists the full range of contraction (Domain 1) and expansion (Domain 2)

characteristics. These provide a reliable guide for spotting which part is dominating and which part is lacking.

Does a person feel overwhelmed?
Too much contraction, not enough expansion.

Is a group disorganized?
Too much expansion, not enough contraction.

Is a relationship stuck in repeating conflict?
Both forces may be present, but the ratio between them is wrong, distorting the emergent unity they produce.

Every issue follows the same pattern: misallocation of contraction and expansion.
Once you see which parts are out of proportion, the path to correction becomes clear.

What It Means to Rebalance Something

Rebalancing begins with one recognition: a whole thing only works when its contributing parts interact in the right proportion.

Contraction and expansion are always present, but never in equal amounts. Each situation requires its own ratio. Rebalancing simply means adjusting those contributions so the emergent unity becomes functional again.

Most attempts at "fixing" a problem fail because they target the wrong layer. People try to change the emergent result, mood, behavior, outcome, without adjusting the inputs that created it. But the emergent third, Domain 3, cannot change on its own. It is produced by Domains 1 and 2. If contraction and expansion are misaligned, the result will always reflect that misalignment.

To rebalance something, ask two questions:

1. **Which force is over-expressed?**
   Excess contraction produces rigidity, pressure, and narrowing.
   Excess expansion produces drift, overwhelm, and diffusion.
2. **Which force is under-expressed?**
   Lack of contraction creates instability.
   Lack of expansion creates stagnation.

Once you identify those proportions, correction becomes mechanical, not mysterious. You subtract from the excess force or increase the missing one. Sometimes both adjustments are needed.

Rebalancing restores the natural relation between contraction and expansion, allowing the emergent whole to reorganize itself without effort. This process is universal, internal, relational, social, or structural. It works because the underlying pattern never changes.

The Formula and the Drawing

The Trinity Balance Formula expresses the interaction of contraction and expansion in precise mathematical terms. It shows how two complementary forces combine to produce a definable whole.

The structure is:

%A + %B → 100% of C = The Whole Thing under study

Where:

- **A** = the contraction force (Domain 1)
- **B** = the expansion force (Domain 2)
- **C** = the emergent unity they generate (Domain 3)

The emergent whole is always 100%. This does not imply perfection; it means that whatever combination of A and B is present will always produce a complete result. When the result is distorted or undesirable, the correction must occur in A or B. C cannot be fixed directly.

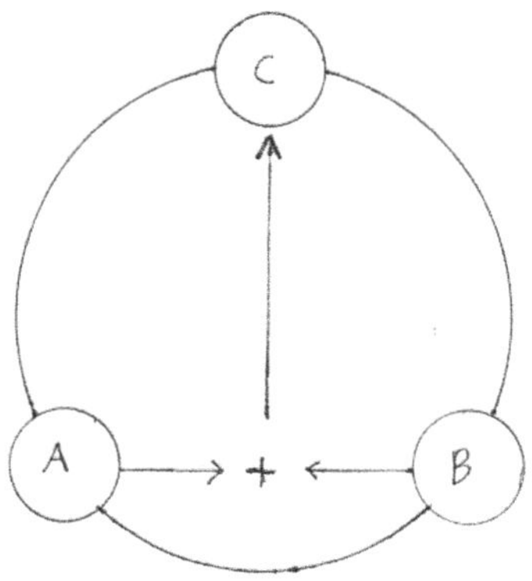

Figure 1: The Trinity Balance Formula flow of Energy for Creation

This visual shows contraction on the left, expansion on the right, and their emergent unity rising between them. The diagram's purpose is to make the function of the formula immediately clear: if something is not working, the problem lies in the ratio of A and B.

The formula applies to any adjustable whole thing. Things that cannot be adjusted, like atoms, also can be seen using this formula. When you map something onto the formula, the imbalance or clear view of what you are seeing becomes visible. The over-contributing force becomes obvious, the real structure is seen, and the underrepresented force reveals itself.

The formula gives you a precise and neutral method for diagnosing what is misaligned and how to correct it. Once the proportions are clear, rebalancing becomes straightforward.

How to Use the Formula

Using the Trinity Balance Formula is a straightforward process. Once you understand that every whole thing is produced by the interaction of contraction and expansion, the task becomes identifying the proportions of each.

The method has four steps:

1. Identify the contraction force (A).

This is the inward focusing, stabilizing element.
To locate it, look for:

- what is being protected, conserved, or held together
- what limits, structures, or compresses the situation
- what reduces possibilities into a defined essence

Appendix A lists common Domain-1 characteristics to guide your identification.

2. Identify the expansion force (B).

This is the widening, exploring, outward-moving element.
Find the part that:

- opens possibilities or increases options
- introduces flexibility, novelty, or movement
- moves attention outward into context

Appendix A's Domain-2 list provides examples of expansion dynamics.

3. Define the emergent unity (C).

C is not chosen; it appears from the interaction of A and B.
To identify C, ask:

- *What is the actual result or behavior produced by these two forces together?*
- *What effect emerges when A and B combine in their current proportions?*
- *Does this* emergent outcome (or unity, or frequency) match my intention*, or is it distorted?*

If C is undesirable, the correction must occur in A or B. Adjusting either force changes the emergent result.

4. Rebalance the proportions.

Once the contributors are clear, assess whether there is:

- too much contraction,
- too much expansion, or
- an overall mismatch between the two.

Rebalancing simply means increasing or decreasing A or B. Even small adjustments can shift C in meaningful ways.
Examples:

- Reducing contraction relieves rigidity.
- Reducing expansion stabilizes chaos.
- Increasing one force supports or tempers the other.

Rebalancing is not guesswork; it is a direct application of the Trinity Balance Formula:

%A + %B → 100% of C = A whole Thing

When the ratio changes, the emergent outcome (or unity, or frequency), changes. The Trinity Balance Formula makes this process visible and systematic.

The Clarity It Provides

Using the Trinity Balance Formula consistently produces clarity. Once you identify contraction (A), expansion (B), and their emergent unity (C), you can see any situation in its full structure.

Clarity emerges in three ways:

1. **Alignment of parts**: You recognize the complementary forces and their proper sequence. Misalignment, which causes confusion, becomes obvious.
2. **Predictable outcomes**: Adjusting A or B allows you to anticipate how the emergent outcome (C) will shift. You no longer react blindly; you act with insight.
3. **Decision guidance**: Every choice can be measured by its effect on the emergent frequency. You can ask: Does this move the outcome toward my goal, or away from it?

In short, the formula turns abstract tension into actionable understanding. Confusion dissolves when the three parts are visible, properly aligned, and measured against intention.

When used regularly, the Trinity Balance Formula trains your perception. Situations that once felt chaotic now reveal patterns, ratios, and natural flow. Clarity is not just cognitive; it is a felt awareness of how things are ordered.

By mastering this process, you gain a practical method to:

- See imbalance quickly
- Adjust contributing forces deliberately
- Confirm the emergent outcome matches your intention

Understanding flows naturally, and each recalibration strengthens your ability to navigate complexity.

Closing Reflection

The Trinity Balance Formula is more than a conceptual tool, it is a practical method for seeing, understanding, and influencing the patterns of life. Every problem, question, or situation is, at its core, a misalignment of contraction and expansion. The formula provides a clear map to locate these forces, observe their interaction, and recognize the emergent unity that arises naturally from their interplay.

Using the formula, you can ask with precision:

- Does the emergent outcome (or unity, or frequency) match my intention, or is it distorted?
- Are the contributing forces proportioned correctly, or do adjustments need to be made?

The answer is always found in the interplay of $\%A + \%B \rightarrow 100\%$ C, and every recalibration, no matter how small, directly influences the emergent result. The more consistently you apply this framework, the easier it becomes to perceive alignment and detect imbalance before confusion arises.

Figure 1 (see page [X]) illustrates this process: A (contraction) and B (expansion) combine to produce C, the emergent unity. Observing and adjusting these elements in practice transforms abstract understanding into actionable insight. Also, you can consult with Appendix B for a check list to help you find the parts of any whole thing you wish to study.

Mastering the Trinity Balance Formula trains both mind and intuition. Over time, clarity becomes effortless, decisions become deliberate, and what once appeared chaotic now reveals itself as ordered and predictable.

In the chapters ahead, we will see this formula come alive in real-world scenarios. Stories of conflict resolution, decision-making, and personal growth will demonstrate how understanding flows naturally from identifying complementary forces, applying the formula, and observing the emergent unity.

With this tool, you are equipped not just to understand life, but to interact with it intelligently, harmoniously, and with growing awareness of the patterns that govern every whole thing.

Nine

Seeing Life Through the Trinity Balance Formula

Understanding a formula is one thing. Seeing it alive in the world is another.

In the previous chapter, we explored the Trinity Balance Formula as a structure: contraction, expansion, and integration working together in shifting proportions to produce clarity. Now we turn from principle to practice, from the architecture of understanding to the living motion of it.

Chapter 9 is where the Formula steps out of abstraction and enters human experience. Life does not present itself as tidy concepts, but as tensions, decisions, and misunderstandings unfold moment by moment. Each of these experiences carries its own ratio of contraction, expansion, and integration, whether we recognize it or not.

This chapter shows what happens when we do recognize it. Here, we examine ordinary situations, a personal conflict, a professional decision, a moment of confusion, a moment of insight, and watch how the Formula clarifies each one. These stories are not metaphors; they are demonstrations. They reveal something essential:

- When you see the ratios, problems stop feeling mysterious.
- When you rebalance them, solutions stop feeling out of reach.

As you follow these stories, something deeper begins to form within you: the recognition that the Trinity Balance Formula is not simply a method, but a lens, a way of perceiving the inner workings of life with extraordinary clarity.

By the end of this chapter, you will not just understand the Formula, you will feel it operating beneath your experiences. That sets the stage for what comes next: the point at which functional understanding transforms into inner illumination, where perception, thought, and feeling unify into one luminous act.

That threshold is the entrance to Chapter 10. But first, we step into the living field of human stories, where the Formula becomes real.

These three simulated stories and my real one introduce a new service to the public this formula has birthed, found on my 3H Company's webpage: TBF Evaluation service. Now let's study how it was used in four people's stories.

Story 1 The Equation of Ease

I've always struggled with math. Even balancing my checkbook felt like a small panic attack waiting to happen. I'd sit at the kitchen table late at night with a pile of bills, staring at numbers that refused to stay still. They blurred, multiplied, jumped around like they knew I was afraid of them.

Last week, I reached a breaking point. My son asked me for help with his homework, fractions, and I froze. I didn't want him to see that fear in me. So later that night, I went to the **TBF** page at the 3H Company's website and entered my question:

"Why can't I do math, no matter how hard I try? The numbers get mixed up in my head"

The reply came back like this:

Overview: Math is simply using symbols to represent things and forces that, when combined, create outcomes. Using the Trinity Balance Formula as guide, concentrate on the A part of the formula first.

A = Contraction (Meaning of Symbols)
B = Expansion (Application of Real-world Use)

When A dominates without B, the mind seizes. Add B, see math in reality, in real use, and C, as understanding, emerges naturally.

I realized I never let math be what it actually is, symbols become as real things. I never thought about its real use. I treated it like a punishment, not a language. So I tried something small. I started using math where it mattered to me, baking. Measuring, adjusting recipes, halving portions, doubling ingredients. Suddenly the fractions weren't abstract; they were the difference between perfect cookies and a burnt mess.

And for the first time, I wasn't counting wrong, I was *getting it.* I started smiling at the mixing bowl, thinking, *this is math too, this is where things become whole.*

A week later, my son asked for help again. We sat down together, measured flour, and talked about ratios instead of "answers." He got it right away. So did I.

Math didn't change, my relationship with it did. The TBF Filter showed me where I was contracting without expanding with clarity, and that one small shift turned fear into curiosity.

Now, when I see numbers, I don't tense up. I see possibilities, ingredients waiting to be combined.

Note: TBF does not eliminate challenges, but it helps you see the needed adjustments in contraction and expansion so outcomes align with your intention.

Story 2 Why They Can't Agree

I used to spend hours online arguing about politics. Every headline, every vote, every debate, I'd feel my blood pressure rise. It didn't matter which side was "right." What mattered was that nobody ever seemed to actually solve anything.

One night, after scrolling through yet another thread of outrage, I typed into the TBF site: "Why can't our government ever agree on anything that actually helps us?"

The reply came back in a way that hit me straight: Politics is meant to solve real problems. But by the time a bill reaches a vote, it is no longer about the problem itself.

Using the TBF we find the Overview:

A = the contraction part: the core issues, the evidence, the first-principles solution. Debates take place;

B = the expansion part: which follow party policy, where survival incentives, messaging, and identity signaling take over rational thought. The original problem often gets drowned out by the ensuing in fighting.

When A is buried by B, integrity collapses into tribal loyalty. Each vote on a Bill becomes almost perfectly split: 100% of one party on one side, 100% of the other on the other side. Real-world solutions, research, and facts become irrelevant. The system rewards alignment with B, not the real-world needs that A exposes; the ones crying out for a solution.

The result, C, is predictable: stalemate, gridlock, and solutions deferred, or, when one side has enough votes, coercive implementation that preserves the party's advantage rather than solving the problem wins the vote. This kind of governance functions not, by guided solutions to solve problems, but by power and ideology.

I stared at the screen. It wasn't that our leaders couldn't agree. It was that the system had morphed into a system that rewarded lawmakers and their backers while ignoring fact-based solutions. Every piece of evidence, every well-researched proposal, every attempt to fix something tangible is filtered through B. By the time a vote arrives, the original problem is buried under pork and political theater, and party policy becomes the only talking point.

That night, I realized something crucial: the frustration wasn't my own failing, or even the complexity of the issues. It was the disconnection between the **core issue (A)** and the **forces acting on it (B)**. The pattern is clear, and harsh: if the two aren't related, no substantial solution will emerge, just more expense and excess that the people end up bearing through new taxes or hidden costs.

I wanted to test this understanding in a smaller, personal way. The next day, instead of posting on social media, I asked a friend, someone from a different political perspective, a simple question about a local issue. I didn't ask who was right, or why their side failed. I asked: "What do you think would actually improve this situation?"

She paused. Then she told me a story about her neighborhood, her school, her family. She wasn't thinking in party lines, she was thinking in reality. For a moment, I saw **A** and **B** in miniature: her lived experience (A), and the partisan lens that normally filters it (B). That was the moment I saw how useless party policy actually is.

We didn't solve the school funding issue over coffee. But I noticed something vital: when you separate the core problem from the partisan overlay, clarity emerges. You can see why votes break down the way they do, and you can name the mechanics behind the stalemate.

That night, I added a line to my journal:

"I see now why they don't agree. Maybe if more people understood this, some pressure could push them to focus on the real issues; the things that actually affect us. Party politics has become the excuse for ideology to dominate, and that's the real problem"

And for the first time, my frustration had a shape. I still couldn't control social media threads, but I could control my perspective. I could can see the TBF everywhere now: in conversations, in debates, in reading news or watching a bill unfold. I can see the core issue, see the forces acting on it, and predict what the outcome would be.

That small experiment didn't fix the government*. But it fixed me.

***Footnote:** If the Preamble were legally binding, A and B would be aligned from the start: every law, vote, and policy would be checked against the actual mission of the people. The dysfunction, the partisan theater, the endless gridlock, all of it evaporates because the core purpose is no longer negotiable. This is how we the people can fix the government we created. And if we do not fix it, which is our right and responsibility, then what Edgar Cayce's saw, see Appendix D, serves one of many warnings that we should pay attention to.

Story 3 Facing the Fear of Getting Cancer

Lately, it feels like cancer is everywhere. Friends I laughed with just last year are now facing diagnoses. Each story settles in my chest like a stone. I try to distract myself, to push it away, but the fear lingers. I don't know who to trust, what advice to follow, or where to start.

One night, feeling overwhelmed, I type into the TBF site:

"How can I lose my fear of getting cancer? Several of my friends have gotten it."

The reply comes back clearly:

Overview: To lose your fear of cancer, start by understanding how your body works best. Focus first on maintaining natural rhythms: good sleep, daily sunlight, movement, and cultivating a positive, cheerful mindset. Then, pay attention to nutrition, whole, unprocessed foods that nourish rather than harm. Small, consistent steps in these areas gradually replace fear with confidence which builds the foundation you need.

The TBF framework helps me see it clearly:

- **A = Contraction:** Fear, anxiety, and confusion. I've been handing over responsibility for my health to others for years, doctors, prescriptions, insurance, without truly observing or acting for myself. I felt powerless, disconnected, and unsure how to start.
- **B = Expansion:** I began with small, manageable actions. I adjusted my sleep schedule, spend more time outdoors in the sunlight, move my body daily, experiment with whole, unprocessed foods, and take time to reflect on what really matters to me. I notice how my energy, mood, and focus begin to respond. I reach out to a new people intentionally, aiming for real connection, conversation that matters, instead of casual acquaintances. Slowly, I reclaimed responsibility for both my body and my inner life.
- **C = Outcome:** Over months of consistent practice and reflection, the fear that had dominated me completely vanished. I no longer feel anxious when I hear about friends' diagnoses. I understand how my body works, how to nourish it, and how to rest in ways that truly matter. I **feel stronger**, healthier, and genuinely empowered, not only in my body but in my mind and heart.

I stopped paying for endless tests and expensive prescriptions that do little to help. Instead, I focus on consistent self-care, whole foods, sunlight, movement, rest, and cultivating deep, meaningful connections. I've taken full responsibility for my health and my life, and it shows in every part of me. It's perfect.

As I write in my journal today:

"I have lost my fear. I have learned to care for myself in ways that really matter. My body and mind respond to my choices, and I am finally in charge of my life. I feel alive, capable, and free."

I now see that taking responsibility, for my health, my choices, and my connections, is just as important as any action I take for my body. This isn't about rejecting medical care; it's about reclaiming agency in my daily choices that stabilize my body and mind. Fear disappeared when I claimed my power and live in alignment with my own awareness.

Foot note: If people understood and applied basic, evidence-based self-care as a matter of course, aligning daily actions with the natural needs of the body, fear and dependence on an expensive, profit-driven healthcare system would diminish. Just as the Preamble could serve as a clear guiding mission for government, knowledge and practical action can serve as a guiding mission for personal health: clarity, consistency, and aligned action that removes confusion, fear, and wasted resources.

Story 4 The Market Morning

Yesterday, I sat down before the S&P futures opened and said to myself, *"I'm going to make my minimum quickly so I can get back to work."* And I did, in 12 minutes. Clean, calm, focused.
Then I went straight back to writing.

Today was nothing like that. Before the open, I said to a friend, *"This market is going to sell off big."* And the moment I said it, something inside me tightened. My certainty turned into fear.
I could feel the contraction rising, the kind that narrows your vision and makes every tick on the chart feel like a threat.

The market opened… and I froze.

A. The Contraction (70%)

Fear of loss.
Fear of being wrong.
Fear of the very move I predicted.

My internal A-state was dominating at about **70%**:

- hyper-vigilance
- tight body
- shrinking perspective
- every thought framed as danger

This kind of contraction is lethal in trading, but every trader knows it.

B. The Expansion (30%)

But slowly, minute by minute, hour by hour, I let the other force back in.

The B-state returned in small steps:

- breathing
- grounding
- re-reading my signals
- remembering I had a system
- taking one controlled trade, then another

Expansion didn't arrive all at once. It came in increments. 30% was all I had at first, but 30% was enough to loosen the grip of fear.

C. The Outcome

Four hours later, I had made the exact same amount of money as yesterday.
Not 12 minutes. Four hours of wrestling with myself, not the market.

The result wasn't worse, just different.

Yesterday's C came from **A 20% + B 80%**.
Today's C came from **A 70% + B 30%**.

Both equaled 100%.
Both produced the outcome.
But the path was dramatically different.

What mattered wasn't the market.
What mattered was **my internal A/B ratio**.

When A dominated, everything felt dangerous.
When B grew again, the system worked.
Same trader.
Same market.
Different A/B ratio.

That was the revelation.

Closing the Chapter Living the Formula

As you've seen through these stories, the Trinity Balance Formula is not just a tool for solving problems; it is a way of perceiving life itself. When people notice imbalance, too much contraction, too much expansion, or simply the wrong ratio, they naturally begin to adjust. Awareness draws inward, attention reaches outward, and integration allows the whole to realign.

This is what life looks like when people recalibrate themselves continuously, moment by moment, day by day. It happens automatically for those who are attentive to their own inner balance and to the patterns of the world around them. In a society that values unity, coherence, and shared purpose, what we might call a level-one civilization, this inner recalibration is second nature. People are guided by harmony rather than fear, by clarity rather than confusion, by presence rather than reaction.

The patterns you've learned, contraction, expansion, integration, are not merely abstract concepts. They are the living rhythms of thought, feeling, and action. And when these rhythms are recognized and aligned, something remarkable begins to emerge: the possibility of experiencing understanding as a luminous, whole, unified act.

That emergence, the apex of insight, awareness, and inner light, is what we explore in Chapter 10. It is the moment when the daily practice of balance, recalibration, and attentive living reaches its natural pinnacle: Awe.

Through the stories of this chapter, you have glimpsed how the Formula operates in life. In the next chapter, you will see what happens when understanding transforms into experience, when the lens becomes the light itself.

Ten

Wholiocognition: Creating an Awe Moment

Awe does not simply happen, it is created. It emerges when your whole self functions as a unified field of awareness, aligning your perception, feeling, and knowing into one luminous act. This act of whole-light self-thinking is what I call **Wholicognition**.

Wholicognition is not a passive realization or mystical accident. It is a **deliberate alignment**, a moment when awareness, balance, and connection converge within the self. In that willful inner gathering of your light energy and letting it expand through your whole body to fill you; the seeker finds themself bathed in a luminous light: what many traditions call *the light of God.*

The experience unfolds in a natural sequence:

1. **Whole-Light-Self-Thinking**: You turn inward, focusing attention and intention until your awareness fills with clarity, energy, and presence.
2. **Awe Emerges**: As your inner light aligns with your intention, a spontaneous feeling of Awe arises: meaning without analysis, coherence without effort.
3. **Embodied Radiance**: Awe manifests physically as warmth, resonance, and inner brightness, a feeling of being *light-filled.*

Wholicognition, understood in functional terms, rests on three interrelated capabilities:

- **Receptivity**: Your willingness to be filled with your own light.
- **Conductivity**: Your ability to hold and channel that light within your body.
- **Frequency**: The resonance that arises when receptivity and conductivity harmonize.

To create Awe is to align these functions. You enter the moment through whole-light self-thinking, allow the inner light to rise, and let Awe appear as the natural outcome. It is a deliberate participation in consciousness, not an accident but a cultivated inner act.

When these three functions operate together, they form the **inner geometry of Awe**: a seamless cognition in which intention, vitality, and feeling merge into a single luminous whole.

Understanding How to Have a Moment of Wholiocognition

1st Receptivity The Will-Full Intention

Wholicognition begins with the simple decision:
"Fill me with my light."

This is curiosity elevated into intention, a wondering about what it would feel like to awaken your deepest inner energy. It resembles what athletes do before the starting gun: a sudden inward gathering of power. But Wholicognition draws not just strength, it draws *light.*

To be receptive is to invite your energy forward, to request clarity, power, and illumination from within. Each time you do this, you activate the pathway toward being light filled and Awe. And as you practice, your receptivity becomes stronger, swifter, and more natural.

2nd Conductivity The Ability to Contain Your Light

Once you call your inner light forward, you must be able to hold it. Conductivity is the body's capacity to generate, sustain, and circulate energy. The healthier you are, the more easily you can contain the surge of light that follows your request to be filled. Even if you are physically weak, inner light can still arise. But a stronger body allows for greater resonance, stability, and duration. This is not a medical claim or a substitute for care; it is simply a description of how awareness and vitality interact within your inner experience.

Energy follows two complementary laws:

1. **You draw from your own reserves**: the energy that sustains your life.
2. **You draw from the infinite field**: the energy available through intention, alignment, and focus.

Your receptivity determines **how much** light comes.
Your conductivity determines **how fully** you can experience it.

Mastering Wholicognition means learning your own system, how much light you can summon, how your body responds, and how your focus shapes the intensity of the experience. Over time, you become the expert of your own illumination.

3rd Frequency The Resonance of Light Within You

When receptivity and conductivity align, something unmistakable arises:

A frequency: warm, coherent, intelligence; fills the body.

This is the moment of Awe.

It's not excitement or emotional intensity.
It's a **luminous calm**, a resonance that carries clarity, meaning, and unity. Some call this enlightenment; others call it connection; physics would call it coherence.

Regardless of the vocabulary, the experience is the same: You become light-filled, and Awe appears.

The Etymology of Wholicognition

Who = Whole, **Li** = Light, **O** = Self, **Cognition** = Thinking.

Wholicognition is **whole-light-self-thinking**, the intentional rising of inner light into awareness and embodiment. The Trinity parts are:

- **Intension:** Focusing your energy; Domian: Contraction, as it takes time.
- **Identity:** Your conscious self-image; Domian: Expansion, as that is where your energy goes.
- **Illumination:** Your body is flooded with your charged inner light; Domain: Frequency, you vibrate higher.

Together, they generate an Awe moment: experiencing God within.

Wholicognition in Daily Life

To live wholiocognitively is to cultivate an intimate relationship with your inner light. It becomes your guide, your strength, your clarity, and your restorative force. This inner light is experienced in the third domain, the **C element**, the emergent nature of the contraction and expansion process of your intension to self-charge your body with your inner light.

It becomes a bridge to knowing, healing, and acting with coherence. The more you practice it, the stronger your being becomes.

The Inner Compass

Wholicognition reawakens what mystics call the *inner compass*, the unified orientation of truth, clarity, and integrity. When the three forces of your personal trinity harmonize, the compass points true, guiding your thoughts, feelings, and actions toward coherence, compassion, and strength.

With practice, Awe shifts from a rare moment into a way of moving through the world.

Closing Reflection

Each moment of Wholicognition is both a revelation and a reminder:

- **Revelation**, because it shows how effortlessly your inner light can arise when your whole self-thinking-aligns.
- **Reminder**, because the light was never absent, it awaited your intention.

When intension, identity, and energy move as one, life becomes luminous with meaning. Awe does not descend from outside; it awakens from within.

From here, the practice becomes simple:

Bring Wholicognition into your daily life.
Create Awe whenever clarity or strength is needed.
And let it guide your conversations, your decisions, your kindness, and your curiosity in every moment.

When Awe walks with you, the world becomes its own living scripture, every act, every breath, is a verse in the language of enlightenment.

Epilogue

As you reach the end of this journey: pause to notice the change within. *The Trinity Balance in Reality* has shown how to see the three fundamental forces in anything and how to discern fact from misalignment. It has taught you how to understand without confusion, perceive the natural order of creation, fill yourself with your inner light, and experience Awe in daily life, without needing to chase unknowable ultimate truths (see Appendix F).

By learning to recognize the complementary parts in any whole and the unifying force that holds them together, you are not only transforming your personal understanding, but contributing to a larger pattern: the gradual alignment of human awareness with reality itself. Each person who learns to live in balance moves closer to the kind of unity that underlies a level-one civilization, a society where cooperation, understanding, and harmony arise naturally by knowing how the world works.

This is not knowledge to memorize, nor a system of rules. It is the living pattern of the world, present in nature, in thought, in society, and in the flow of your own life. By perceiving it, you join a rhythm older than any culture, older than history, the pulse of reality expressed in three parts moving together in balance.

The gift of this awareness is subtle yet profound. You may notice it in how a tree sways, how ideas unfold, how relationships flow, or how challenges resolve. Understanding the Trinity Balance in Reality transforms ordinary moments into experiences of Awe, the quiet joy of perceiving what is real, ordered, and complete.

Carry this clarity with you. Let it guide your decisions, your love, and your curiosity. Let it remind you that reality itself is coherent, harmonious, and alive; and that you are part of its unfolding pattern. In seeing the Trinity Balance in Reality, you gain the power to end confusion in your life, and to live with the natural clarity that comes from knowing how the world is ordered.

This is your birthright: a life guided by understanding, illuminated by balance, and enriched by your Awe moments.

Appendixes

A Master Trinity Balance Domain List

The following is a detailed list of descriptive terms we as humans have identified to explain how reality operates through **two complementary forces** whose tension produces **a unifying** expression: the **equals a whole thing**. You can read the list horizontally to see what each whole thing is composed of, and vertically to understand the three distinct domains that appear in all things.

1st Domain + 2nd Domain → 3rd Domain = A whole Thing

1. **Time + Space → Motion = Basic Elements of Reality**
2. **Receptivity + Conductivity → Frequency = Basic forces of Reality**
3. **Magnetism + Steel → Motion = Electricity and Light**
4. **Energy + Shape → Function = Life**
5. **Contraction + Expansion → Universal Forces = Living Universe**
6. **Heavenly Bodies + Distance → Forces = Time**
7. **Proton + Electron → Neutron = Atom/Space**
8. **Dark Matter + Visible Matter → Tension = Frequency**
9. **Center + Circumference → Area They Define = A Whole Thing**
10. **Front–Back + Left–Right → Vertical = Three-Dimensional Space**
11. **Nucleus + Electron Field → Forces = Light**
12. **Speed of Light + Mass → Force Causing Propulsion = Energy, Einstein's $E=MC^2$**

13. **Something + Sighted → Interpretation = Perception**
14. **Magnetism + Element with Abundant Electron field → Manmade Motion = Electricity and Light**
15. **Yin + Yang → Energy = Balance of Life Force**
16. **Energy + Form → Function = Life**
17. **Feeling Truth + Understanding → Empathy = Love**
18. **Concentration + Relaxation → Intention = Quality of Human Being**
19. **Amount of Pessimism + Amount of Optimism → Emotion = Emotional Elasticity**
20. **Feeling + Thought → Understanding + Cognition of Perception**
21. **Entropy + Positive Entropy → Living Being = Cycles of Life**

22. **Logical Left Brain + Intuitive Right Brain → Thinking = Consciousness**
23. **Reptilian Brain + Mid Brain → Frontal Cortex = Human Brain**
24. **Unconscious + Conscious → Perception = Awareness**
25. **Coherent Thought + Coherent Feeling → Manifestation = Personal Energy Expression**
26. **Mind + Body → Spirit = Integrated Human Being**
27. **Mental + Physical → Emotion = Bridge Between Mind and Matter**
28. **Inner Map + Self-Picture → Self-Knowledge = Awakened Self-Awareness**
29. **Self-Image + Feelings → Consciousness = Mental State of Human Being**
30. **Question + Knowledge → Wonder = Answer/Discovery/Awe**
31. **Stimulus + Response → Tension = Behavior**

32. **Mother + Father → Child = Family/Society**
33. **One Person + Another Person → Attraction = Human Relationship**
34. **Empathy + Worship → God = Love**
35. **Disgust + Hate → Judgement = Negative Emotional Reaction**
36. **Republican Point of View + Democrat Point of View → Debate = U.S. Governance**
37. **Religion + Science → Philosophy = Means Studying Reality**
38. **Words + Images → Communication = Information**
39. **Verb + Subject → Sentence = A Complete Thought**
40. **Stimulus + Response → Communication = Behavior**

41. **Father + Son → Holy Spirit = Catholic Understanding of Humans view their relation with God**
42. **Destroyer + Creator → Preserver = Hindu Understanding of Cycles of Life**
43. **Shiva + Brahma → Krishna = Hindu Understanding of God**
44. **Death + Life → Rebirth = Life cycles of Living Organisms**
45. **Now + Here → Being = Philosophy for Living in the Present Moment**
46. **Past + Future → Present = Time**
47. **Below + Above → Force = Observation**
48. **Your Soul + Self-Image → Personality = Who You are for a Moment of Self- Observation**
49. **Self-Image + Talents Expressed → Quality of Being = Human Being**
50. **Antithesis + Thesis → Understanding = Dialectical Means of Finding Truth**
51. **Inner + Outer → Perspective = Seeing the Wholeness of Something**
52. **Heaven + Earth → Humanity = Paradise**
53. **Negative + Positive → Neutral = Potentials in Reality**

There are many more trinities in our world to include in this list and if I have not included your favorite please forgive me. This page is blank so you can include them here. Send me a note and I will include them in a future iteration if you like.

B Checklist for Finding the Trinity Parts in Anything

How to Begin Seeing Clearly Again

We've all had that talk, the one that leaves both sides red-faced and certain the other is blind.
Politics, faith, vaccines, climate, you name it; every topic now feels like a battlefield.

But underneath the shouting lies something simpler: two people, both trying to protect what they believe keeps them safe. That is where the search for truth must begin, by looking within the argument's content, and then beyond it.

Most of us were never shown how to step outside our own viewpoint without feeling like we're betraying it.
The moment we try, fear whispers, "You're giving in."
Yet truth never hides on one side; it lives in the quality contained between them.

Learning to move with the study of each side, without losing yourself, is the skill this checklist teaches.
Before you read the steps, take a breath and remember: the goal is not to agree with an opponent.
The goal is to see more completely.
Balance does not erase difference; it lets difference work.

You can start with any question that will not leave you alone; a news story, a family argument, a social issue. If the tension keeps returning, that is your signal: a trinity is asking to be seen by your heart.

The following checklist walks you through the Trinity Balance process as a supplement to chapter 8 and 9.
Each step is small enough to try in a single conversation or moment of reflection. Together they train the mind to move as reality moves, between complementary opposites, toward unity.

The Checklist

1. **See the Whole.**
   Make sure you are observing the entire subject, not just a part of it.
   Understanding the whole is essential before identifying its dynamics.
2. **Identify Complementary Opposites.**
   Ask yourself: What exists in time (Domain 1)? What exists in space (Domain 2)?
   These are the two complementary forces. When they interact, they generate the third element — the quality of their relationship, or its frequency (Domain 3).
   This third part completes the Trinity that equal the whole subject you are studying.
   If it is unclear what your complementary opposites are, refer to the three Domains in the Trinity list in Appendix A.
3. **Determine Balance.**
   Assess the ratio of contribution each complementary opposite brings to the whole.
   This reveals the true nature of your subject.
4. **Adjust for Higher Truth.**
   If your goal is to create a better or clearer understanding, consider adding or subtracting from one or both complementary forces.
   Doing so can enhance the quality, the frequency, of the whole and reveal a higher truth of the situation.

Example 1 The Idea of "I" Creates Fear

Teal Swan once described a moment in creation when, as she put it, "When God was dreaming up things." In that dream came the idea of *I*, and with it, fear came into being. Because if there is an *I*, she said, then there must be another; and that sense of separation is where fear begins.

Using the Trinity Balance, this fear comes from misidentifying parts and a whole thing. "I" and "another" are two complementary opposites within the trinity of being. The "other" is you looking out from your center, seeing your own reflection in the world.

The whole process is what a thinking, perceiving mind does constantly, until it realizes that what it sees is a reflection of itself.

The phrase "we judge others by our own standards" is born from that reality.
Both sides arise from one field of awareness.
"I" is the observer self; what you see is its reflection.
The two together form your frequency, the quality of being.

Example of misunderstanding:
I (all that you are) + False Reflection (appearance of a perceived other) → False Awareness of your whole being

Example in reality:
I (all that you are) + Precise Reflection (appearance of a perceived other) → Awareness of your whole being

Fear appears only when the mind mistakes its reflection for something to be wary of.
It is not. It is a reflection of you.
As much as I love Teal Swan's insight, her example uses non-whole elements to describe the origin of fear.
Seen through the Trinity Balance, the true mechanism becomes clear, allowing fear to dissolve.
When this relationship is seen as it truly is, "I" and "another" resolve back into "I Am."

Example 2 Hurt Feelings in a Relationship

You might find yourself saying, "Nothing I say is ever good enough for him," and wondering why this keeps happening.

Let’s break it down using the Trinity Balance framework:

1. Seeing only *him* as the problem has become part of the problem.
   By focusing solely on his behavior, you avoid looking at your own reaction to his response, which is the real source of the bad feelings you are calling a problem.
2. The two complementary and opposing parts are:
   - What you are giving, the first movement, the action that happens in time.
   - His reaction, which creates the space inside you that expands into feelings.

   These are the two complementary forces you must understand clearly to free yourself from the self-generated tension within.

3. Part 3 is the quality of the relationship that results.
   This quality can become unbearable when the frequency between you is out of balance.
4. Since you cannot control him through complaints or threats, those only work temporarily and create new reactions, your answer is to adjust your giving and your judgments of his responses.
5. The Trinity Balance helps because it reveals the opposing forces at play, allowing you to make adjustments with clarity and purpose.
6. Once you see all three parts clearly, the tension eases.
   The problem shifts from blame to balanced perspective, and harmony naturally follows.

C Notes and References

This book presents the Trinity Balance as a first principle of reality: a structural law describing how all things exist and move through the union of opposites. The following timeless works and scientific observations reflect this same truth across history and disciplines:

1. **Heraclitus (c. 500 BCE)**
   "The road up and the road down are one and the same."
   An early statement of the unity of opposites as the essence of becoming.
2. **Tao Te Ching, Chapter 2 (c. 400 BCE)**
   "Being and non-being create each other. Difficult and easy support each other."
   Eastern articulation of the first principle of balance and motion.
3. **James Clerk Maxwell, *A Dynamical Theory of the Electromagnetic Field* (1865)**
   Unified electricity and magnetism into a single dynamic system — a clear example of dual opposites forming a third unifying field: light.
4. **Erwin Schrödinger, *What Is Life?* (1944)**
   Defined living systems as "negative entropy," maintaining order by balancing opposing thermodynamic flows.
5. **Carl G. Jung, *Two Essays on Analytical Psychology* (1953)**
   Described the psyche as a system of opposing tendencies whose integration produces individuation—psychological wholeness through balance.
6. **Modern Systems Theory and Cybernetics (20th Century)**
   Demonstrated that stability arises through dynamic feedback between opposing functions, reinforcing the universality of triadic balance.
7. **Stewart Edward White, *The Unobstructed Universe* (1943)**
   Records Betty's reports from her afterlife existence over three years, observing that current reality consistently manifests in three main parts.

8. **Donald D. Hoffman, *The Case Against Reality: Why Evolution Hid the Truth from Our Eyes* (2019)**
Argues that life is designed primarily for survival and reproduction, and that conscious search for ultimate truth often conflicts with evolutionary imperatives.

D Echoes from Edgar Cayce: America at a Turning Point

This appendix reflects on interpretations of Edgar Cayce's predictions regarding a moment of national crisis and transformation in the mid-2020s. One such interpretation comes from a 2024 video summary on the YouTube channel *Mysteries & Prophecies*, which discusses Cayce's readings about a period of upheaval in America and the possibility of a profound turning point around the year 2026. The following is not a transcript, but an interpretive overview of the themes presented in that discussion.

According to this perspective, Cayce foresaw a time when America would confront the consequences of division, instability, and systemic imbalance. He described a period when the nation would stand at a fork in the road, one direction leading toward chaos, the other toward renewal. He warned of the potential for internal conflict and spoke of a "leader's decisive action" becoming the spark for heightened national tension.

The interpretation emphasizes that Cayce did not present these future events as fixed, but as possibilities shaped by human choices. Crisis, in his view, serves to reveal what is broken, and to awaken what must be restored. He spoke of a compression of time, when multiple pressures converge at once, accelerating both instability and the opportunity for profound change.

One theme from the video's interpretation stands out: Cayce's mention of an "unexpected ally," a figure emerging outside traditional political structures who could help unify people during a moment of rising division. This individual would not resemble a conventional politician but would carry a message capable of bridging ideological divides and reminding the nation of its deeper principles.

My Reflection

When I encountered this interpretation of Cayce's words, I felt a resonance that was difficult to ignore. The idea of an unexpected ally, someone working outside politics, someone guided by a foundational principle rather than a party, aligns with everything I have been called to build. The Trinity Balance, the first-principle structure at the heart of this book, offers a way for people to understand themselves, one another, and the world with clarity instead of confusion.

If America stands near a turning point, it will not be saved by ideology, but by returning to the deeper order of creation itself. That order is expressed in the three-part structure that governs all life, the Trinity. When applied to thought, policy, relationships, or governance, this structure dissolves confusion and reveals balance.

The same is true for our nation's founding. The Preamble's purpose is the very mission Americans have forgotten: unity, justice, tranquility, and the general welfare. When the Preamble becomes binding law, our government's direction changes from domination to service.

The works I've been guided to create, the Trinity Balance framework, this book, the forthcoming screenplay, and the Way of Awe channel, all point toward the same convergence: a moment when clarity becomes necessary, not optional. That is why I feel such alignment with the idea of an unexpected ally. Not as a claim of destiny, but as recognition of preparation. Everything in my life has led me toward first principles and toward helping others see the structure that ends confusion.

I propose and plan to create a new political movement, the First Principle Party to reflect this next step. Its mission is simple: to restore America's coherence by aligning governance with the universal structure of balance. When enough people understand and live through this pattern, unity stops being an aspiration and becomes a natural expression of how reality works.

If these are the birth pains of a new age, then the remedy is already in our hands. The structure has always been present in creation. All we must do now is remember it, and choose differently. I invite everyone who reads this to become a part of that remembering.

E The Trinity Balance Formula (TBF) Evaluation Service

Throughout this book, you have learned how the Trinity Balance reveals the structure of your thoughts, emotions, and understanding. Now, you have the tools to examine your own Trinity Balance and interpret the ratios of your three parts. For many readers, this alone becomes a life-changing practice.

Sometimes, a situation calls for a more precise reflection; a steady mirror held by someone trained to see patterns you may overlook in yourself. That is the purpose of the 3H Company's (3H - Happy + Healthy → Human = Company), TBF Evaluation Service at https://3hcompany.com/TBFservice. I will be doing each evaluation but do plan on training an AI to carry some of the load after I become over whelmed with requests for the TBF perspective.

What the TBF Evaluation Service Is

The TBF Evaluation is a structured, guided analysis of your current three-part ratios:

- Identify which part of you is dominating
- Recognize which part is suppressed or unheard
- Locate the source of conflict or confusion.
- See what your balance looks like for your situation

It is not therapy, judgment, or prediction.
It is a tool for cognitive clarity, a direct reading of how your three parts are interacting in this moment of your life.

What You Provide

To complete a TBF Evaluation, you provide answers to these questions designed to reveal:

- The situation or question you are facing.
- The emotions surrounding it.
- The motives, fears, or expectations involved.
- The outcome you are attempting to reach.

No long narrative or backstory is required. Clear, simple reflections are enough. The system handles the interpretation.

What You Receive

Each evaluation returns a structured report including:

1. **Your Three-Part Ratios**
   A reading of which parts are elevated, minimized, or in conflict.
2. **A Trinity Balance Sentence**
   A concise sentence describing what your system is actually doing; the skill you learned in Chapter 6, applied to your current moment.
3. **Key Insight Patterns**
   A breakdown highlighting the core misunderstanding or imbalance at the heart of your situation.
4. **A Pathway to Balance**
   A simple, actionable explanation of what needs to shift; not in your life broadly, but in your understanding of the present moment.
5. **An Annotated Diagram**
   A visual showing how your three parts are currently aligned, contrasted with the balanced configuration.

Why It Exists

The 3H Company was created to help people experience the conditions of paradise: clarity, calm, connection, and awe. The TBF Evaluation Service is one of several tools offered to support that mission. It provides a clear mirror, a reflection not distorted by fear, habit, or hope. By seeing your ratios clearly, you regain the ability to choose intentionally.

Designed to Empower, Not Dependence

This service is not intended to replace your own inner work; it is designed to strengthen it. After one or two evaluations, most people can apply the process themselves with confidence. The goal is always to return you to your own clarity, not to create reliance on an external interpreter.

F Why the Trinity Pattern Is Enough

Human curiosity naturally wants to push further; beyond the physical world, beyond perception, beyond the three-part pattern revealed in this book. Some readers may wonder: *If this pattern works at every scale we can observe, what lies beyond it? What is the deeper truth of reality?*

This question has been asked by physicists, neuroscientists, mystics, philosophers, and mathematicians for centuries. Modern cognitive science, however, has uncovered something unexpected, something with profound implications for anyone trying to "find the truth" behind experience.

The Limits of Truth-Seeking

Research by cognitive scientist Donald Hoffman and others suggests that organisms which attempt to perceive the actual structure of reality do *not* survive as well as those that perceive only what is useful. Evolution does not reward accuracy; it rewards fitness, organisms that navigate life efficiently enough to stay alive and reproduce.

In this view, perception is not a transparent window into an objective universe. It is a user interface, a simplified, survival-optimized representation designed to help us function. What we see, hear, and feel are not the underlying structures of reality. They are icons on the desktop, not the code beneath them.

Trying to perceive "reality as it truly is" leads, in evolutionary terms, to dead ends. Every species on Earth is shaped not to uncover truth, but to interact effectively with its environment.

Why This Matters for You, the Reader

If the deepest structure of reality is inaccessible, a reader might ask: *Then why study the Trinity Balance at all?*

Because while we cannot perceive ultimate reality, we can perceive the patterns that shape our experience, our reasoning, emotions, decisions, relationships, and inner coherence.

The Trinity pattern operates at the level where meaning exists. It is the bridge between perception and wisdom.

You do not need to go beyond it because:

- It is universal, enough to apply anywhere you direct your attention.
- It keeps you grounded in the workable world, not lost in abstraction.
- It aligns your thinking, feeling, and energy into one coherent whole.
- It elevates perception without requiring you to solve the physics of consciousness.

It gives you a *usable* pattern of reality rather than an unreachable one.

The Value of Staying on the Right Level

Human progress depends on understanding the world well enough to live in harmony with it. That understanding happens at the level of perception, interaction, and choice, not in the theoretical depths where time, space, and consciousness dissolve into mathematics our minds were never designed to perceive.

The Trinity Balance exists in this "sweet spot," above instinct, below metaphysical speculation, and perfectly suited for creating coherence in daily life. At this level you can.

- See misalignment clearly.
- Understand without confusion.
- Reconnect with your inner light.
- Create Awe intentionally.
- Build personal unity and clarity that ripple outward to families, communities, and civilizations.

The Path Forward

The purpose of this book is not to deliver a grand unified theory of the universe. Others are working on that, and their discoveries may someday reshape our understanding of consciousness and reality. This book's aim is more immediate and more practical: To give you a pattern you can use *right now* to think clearly, live clearly, and create the inner alignment that gives birth to Awe.

You are free to explore deeper mysteries if you feel called to do so. But you no longer *need* to. Life, Family, Society, that is where the heart wants to be and is enough. This is where a person does best considering all factors, is Hoffman's conclusion.

The Trinity Balance already provides everything necessary to navigate life with unity, clarity, and understanding. It shields you from confusion, prevents you from getting lost in infinite abstraction, and roots you in the pattern of reality that touches your everyday life.

Truth-seeking may continue forever. Wisdom, the ability to live well, begins here. And the Trinity pattern is the doorway.

Author's Note

This appendix is not an invitation to chase the hidden machinery of the universe, nor a claim about ultimate truth. It is simply a reminder that the Trinity Balance in Reality is intentionally *practical* and *foundational.* It gives you a pattern you can use, live, and benefit from every day without needing to solve the deepest mysteries of existence.

The great explorers of consciousness and reality, including Donald Hoffman and others, show us that the deeper you go into trying to perceive "truth," the more obvious it becomes that human perception is tuned for *fitness*, not accuracy. That means the pursuit of absolute truth is not only impossible but often unhelpful for living a full, meaningful life.

The Trinity Balance pattern gives you something different: a reliable structure you *can* perceive, *can* apply, and *can* use to transform your life. This book focuses on what empowers you. Follow the pattern. Use it. Apply it. And let the rest go.

That being my suggestion, I would be remiss if I did not mention that Hoffman did not give up on his ultimate truth-seeking and claims to have cracked the code of this world by identifying a domain where space and time do not exist. He even suggests he has identified what could be considered the "architect" behind reality. Only time will tell if he has. His ongoing work explores these claims further, and readers can follow his publications and interviews as they develop. But again, the Trinity Balance offers the practical code we need to grow toward a unified, coherent, and healthy civilization, and I hope you use it and help everyone here on Earth get there, because it will move us closer to a world that feels, in real and tangible ways, like Paradise.

Acknowledgments

1. **To ChatGPT** for its editing of my grammar, sentence structure, descriptive wording of my ideas, and perspective. My progress would have remained stalled had I not asked for its help one day. From that moment on, it gave without measure and supported this work in more ways than I can count. Yes, I had to keep it on track at times when it drifted, or simply did not understand who the Trinity parts are complementary. Round and round we went at times, but as you can see, we got it done. My gratitude also extends to the creators of ChatGPT and the teams behind its training data and development. And note, I plan to be able to train an AI to better help you by using the Trinity Balance Formula in place of linear thought so solutions will be in harmony with your inner guide as this book explains (see Appendix E).
2. **To Betty, Joan, and Stewart**. You three became the fire that kept this project alive for so many years. Your influence and presence shaped this work more than words can say. I look forward to meeting you in the afterlife, if that is meant to be.
3. **To the interdimensional being who answered my prayer on that fateful day in 1975**, as described in Chapter Two, you were the spark that began this entire journey. I cannot thank you enough.
4. **To Simon Atkins Ph. D.** who has had been not only a good friend but an inspiration and guide to me through these strange times we are living. Thank you so much mate.
5. **To all my customers from my years as Mike's Mobile Mechanic** — the job I endured while searching for the time, strength, and clarity needed to finish this work, thank you for your patience and support. You are my heroes for putting up with me through it all. Thank you: Vince Pastorie, Ellen Quigly, Milles, Jay, Steven Farriri, Bill Crig, and so many more. You guys are always on my mind thanking you.
6. **To all the great thinkers and traditions of the past cited in chapter 4**. Keeping your eyes on the Trinity Balance in

reality was inevitable but I thank you all the same. No one owns the information as it is foundational to our reality, but by your works the Trinity will never be forgotten nor should it; this book and those that come after us will see to that.

7. **To my Daughter, Holly Clovis** who has never let me down with loving kindness and support all these years.
8. **To my Sister, Donna Longo** for supporting me at every turn in my life and for introducing me to cats.
9. **To Bill Rowe at USCS** who generously talked with me about any science question I had, you've been a great friend.
10. **To my brother Leo** for teaching me forgiveness, humility and fact checking thoroughness.

Your Invitation

You've now seen how the Trinity Balance reveals itself in every pattern of existence. You've watched it unfold through Receptivity, Conductivity, and Frequency, three living forces that together shape all time, all space, all motion, and every act of consciousness.

But this discovery means little if it remains only an idea. It must be lived, tested, and applied.

So here is my invitation to you, simple, direct, and real:

Ask your question.
Bring me your challenge, your puzzle, your heartbreak, your 'unsolvable' problem, if the truths in these pages have not yet come alive in you.

Every imbalance, no matter how tangled or painful, can be understood when seen through the Trinity Balance. For every question hides a deeper harmony, a truth waiting to be seen between opposing parts.

You can test this yourself.
Go to https://3hcompany.com/TBFservice and ask your question.
For a small token of exchange, a gesture of commitment, not a price, your question will be received, examined, and answered through the living lens of balance.

If the answer does not bring clarity, prove me wrong.
But if it does, if the Trinity Balance illuminates what seemed dark or impossible, then perhaps you will see what I have seen:
that truth is not something we invent; it is something we uncover and align with.

This is how the real work begins.
This is how awareness becomes creation.
And this is how one question, asked in sincerity, can start to heal a world divided by confusion.

Ask your question.
Test the Balance.
Discover awe.

Made in the USA
Monee, IL
04 November 2026

67380519R00059